Gilvete Silvania Wolff Lírio
Adriano M. Souza

Evaluating service satisfaction using multivariate analysis

Gilvete Silvania Wolff Lírio
Adriano M. Souza

Evaluating service satisfaction using multivariate analysis

A case study in a telecommunications company

ScienciaScripts

Imprint

Cover image: www.ingimage.com

This book is a translation from the original published under ISBN 978-3-330-77071-3.

Publisher:
Sciencia Scripts
is a trademark of
Dodo Books Indian Ocean Ltd. and OmniScriptum S.R.L publishing group

120 High Road, East Finchley, London, N2 9ED, United Kingdom
Str. Armeneasca 28/1, office 1, Chisinau MD-2012, Republic of Moldova, Europe
Managing Directors: Ieva Konstantinova, Victoria Ursu
info@omniscriptum.com

Printed at: see last page
ISBN: 978-620-8-63359-2

SUMMARY

Statistics is the art and science of what is difficult to perceive at first, but which once revealed becomes obvious.

(Dr. Victor Kane - *Ford Motor Company*)

1. SERVICE QUALITY AND CUSTOMER SATISFACTION

In today's context, public and private organizations are increasingly concerned about the quality of their products and services, which are the differentiators between companies. The relationship between company and client is the main focus of decisions, in other words, clients are increasingly demanding in their search for certified or accredited services.

Concern about the quality of products and services goes beyond the boundaries of Brazilian companies. This is measured among American companies by the Malcom Baldrige National Quality Award (1990), which measures customer reaction to the products on offer and is awarded annually to American companies that demonstrate high standards of business practice. This award includes seven categories, the most important of which is customer satisfaction.

Within this category, companies are judged on their knowledge of customer needs and expectations, the management of their relationship with customers, the commitments they make, the methods used to determine customer satisfaction, the results of customer satisfaction and, finally, comparisons of the degree of customer satisfaction with that of their competitors (Hayes, 2001).

According to research carried out by SEBRAE (Brazilian Micro and Small Business Support Service), the vast majority of small businesses (79%) have shown an interest in learning about and applying processes that increase the quality of their services or products.

A service or a product has quality when it meets the real needs and expectations of customers, in all the tasks carried out in the company. It therefore requires serious work, the success of which will only be achieved if all the people involved in the process work together, aiming for continuous improvement.

Measuring customer satisfaction and evaluating the degree of customer satisfaction is still an important topic of research today, as it must use statistical techniques that capture the simultaneous interrelationship between the variables evaluated.

Nowadays, companies not only rely on their employees to improve quality standards, but also on specialized consultants as well as techniques and methodologies, with the main objective being to contribute to increasing quality, which, according to Montgomery (1985), is the extent to which products or services meet the requirements of the people who use them.

There are also two types of quality: design quality and performance quality. Design quality reflects the extent to which a product or service possesses an intended characteristic, while performance quality meets design expectations.

Evaluating quality and customer satisfaction is more appropriate for companies in the service sector, as it helps them to focus their attention on customers and how they receive services from the company.

Customer satisfaction surveys and literature on the subject, such as Kotler (1996) and Mattar (1993), show that the main aspects that should be taken into account when evaluating a service are: promptness, availability and professionalism. Using these three items, the questionnaire was drawn up, which consisted of 14 questions, in which statistical techniques were applied to reduce the number of variables and then analyze the variables considered most relevant to the customer choosing a particular service.

The general satisfaction of service quality has observable characteristics, i.e. a customer can smile or make good comments about a service received, which can be called "customer satisfaction", or do the reverse process, although satisfaction or dissatisfaction are not opposites. According to Juran (1990), satisfaction with a product or service comes from its characteristics and is the reason why customers buy the product or service. Dissatisfaction, on the other hand, stems from non-conformities and is therefore the reason why customers complain.

According to Bateson & Hoffman (2001), *apud* Friedrich, Quadros & Viegas (2003), a very simple process for evaluating after-sales satisfaction would be to replace the various services or products offered by the company with two columns: expectation (E) and perception (P), for each attribute, and, in the end, subtract one from the other (P-E). With the appropriate weighting, this gives the result of the degree of satisfaction on the part of the consumer.

The article published in the magazine ESPEM (Escola Superior de Propaganda e Marketing, 2003), entitled "Measuring consumer satisfaction in Porto Alegre hotels", by Friedrich, Quadros and Viegas, points out that the cost of new customers is increasingly expensive, which makes keeping old customers more economical, i.e. the level of satisfaction consequently the level of retention of these customers by companies.

That's why it's important to carry out a research process that provides information considered relevant to maintaining and growing competitive advantages over potential customers, so that managers can make decisions.

Figure 01 shows the main benefits of an effective service quality information system.

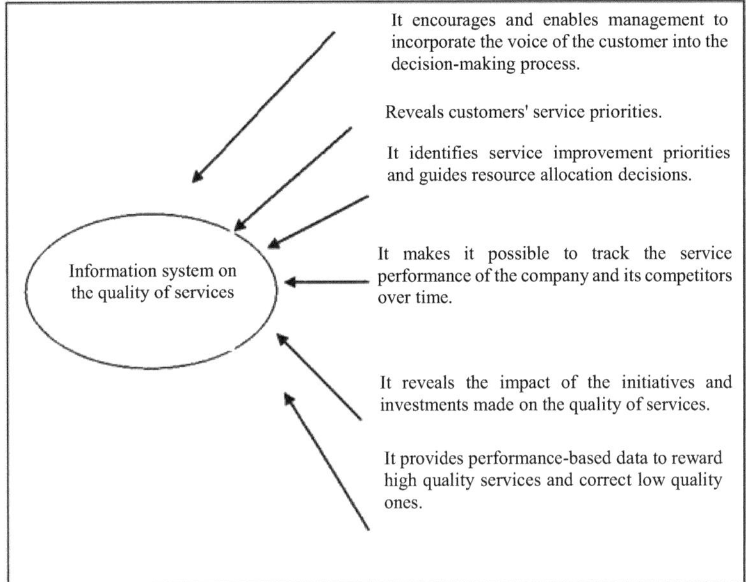

SOURCE: Berry (1992), p. 35.

FIGURE 01- Main Benefits of an Effective Service Quality Information System

Therefore, if the manager carries out a continuous survey of satisfaction levels and service quality, the detection of possible problems that may arise will be quick enough for correlations to occur, adjusting the process or service offered based on the demands of potential consumers.

Lovelock & Wright (2002) *apud* Friedrich, Quadros & Viegas (2003), list six benefits of customer satisfaction, as shown in Figure 02.

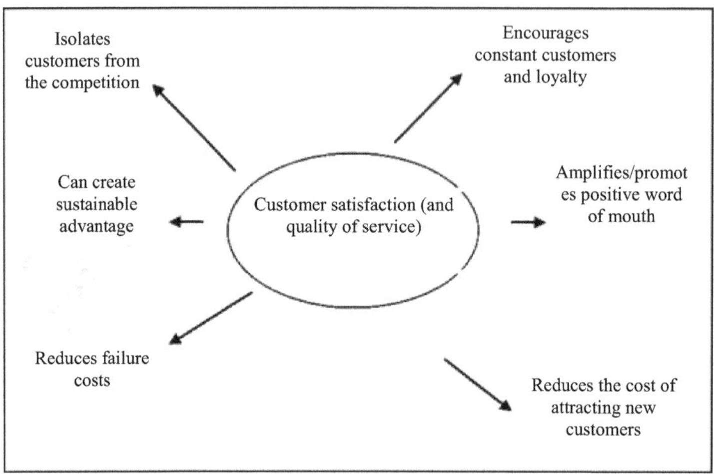

SOURCE: Lovelock & Wright, (2002), p 116.

FIGURE 02- Benefits of Customer Satisfaction and Service Quality

The level of satisfaction can be measured using statistical indices or more advanced techniques that produce better results.

Few research studies in this area are related to the multivariate analysis technique used in this study. Only a few studies were found, but these were more related to the area of *marketing* research, such as the work done by Latif (1994), in which the author applied the technique of factor analysis to solve a real *marketing* research problem in a supermarket. This study investigated the criteria people use to choose the supermarkets they shop in.

In Malhotra (2001), you can also find some studies applied to the *marketing* area. These include a customer loyalty project for a department store, in which 21 different lifestyles were assigned and analyzed factorially in order to determine the fundamental factors of lifestyles. At the American company Burke, a recent project was designed to reduce the number of initial variables (16 questions), and a principal component analysis was carried out to check which groups of questions were highly correlated, helping them to use the answers to better interpret the results of satisfaction surveys carried out by the company.

It is then up to the company to provide people with the knowledge they need to continuously improve services, so that they are responsible for the quality of their own work, thus facilitating relations between the company and the customer, and making them motivating elements and agents responsible for the success of ventures and ensuring that the company survives in terms of quality, productivity and service.

Therefore, the quality of their services is often difficult to gauge, because most of the time there is no *feedback* from the client to the company on the service that has been offered, nor on the return that has been obtained from the investment made.

In this way, the company must have the flexibility to change and adapt to customer needs, the empathy to provide differentiated service to each customer, the reliability to provide services as promised and the speed of customer service. All these items put together, combined with good human resource performance, will mean that the company will be able to satisfy its customers more quickly after receiving a given service.

Statistical tools are used with the aim of reducing and controlling the uncertainties involved in decision-making situations, as well as helping to reduce the variability of the processes analyzed (Werkema, 1995).

Among the statistical tools used were sampling, descriptive statistics, data segmentation by cross-referencing variables, and multivariate statistics.

A tool of fundamental importance for data collection, in addition to questionnaires, is the attitude measurement scale, which is used to create a quantitative variable from a multivariate frequency distribution. Attitude measurement scales include the *Thurstone* scale, the *Guttman* scale and the *Likert* scale.

In order to prepare the questionnaire to measure the level of satisfaction, we chose to use the *Likert* scale, because this type of questionnaire allows customers to answer each item in varying degrees, which has a neutral point as if it were zero, which represents the center of the scale. From a statistical point of view, this variety of degrees is more reliable when analyzing the data, as scales with five options are more reliable than those with only two, like the *checklist* type. In addition, the use of the *Likert* scale allows the percentage of positive or negative responses to a given question to be determined, as shown in Figure 03.

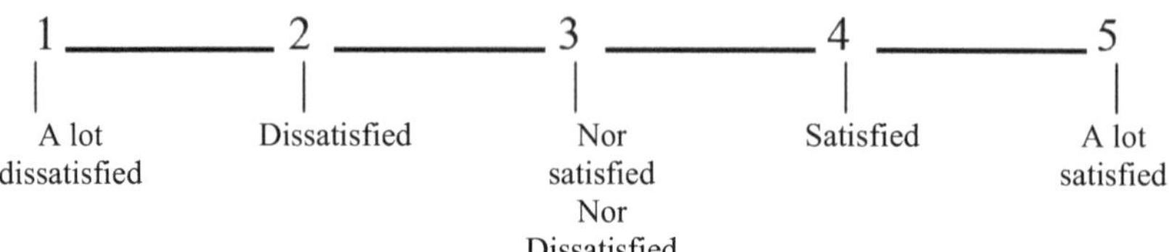

FIGURE 03 - Schematic of the five-point *Likert* scale

The concern for quality has been demonstrated by companies since the 1950s, first being implemented in industries and then in service companies, which recognized the complexity and differentiation between products, goods and services.

Thus, in order to establish techniques for the pursuit of service quality, it is necessary to use statistical methods as effective tools in the process of analyzing and understanding what has been collected.

The use of Multivariate Analysis (MA) will be useful for discovering regularities in the behavior of the variables collected, including determining how and when two or more groups differ in their profile in multivariate terms.

For Pereira (1999), AM involves a wide range of statistical and mathematical concepts and is, strictly speaking, any analytical approach that considers the behavior of many variables simultaneously. For Kendall (1963), the values of the different variables must be obtained on the same individuals, and they must be interdependent and considered simultaneously. The technique has been used successfully in the most diverse areas of knowledge; some examples of application are presented by Hair Jr. *et al.* (1995).

Based on this, multivariate methods were used, including Principal Component Analysis (PCA), Factor Analysis (FA) and Cluster Analysis (AA), also known as *Cluster* Analysis (CA), as they are methods for summarizing data sets, They are methods of summarizing data sets which have various characteristics, but which, by means of a few parameters, adequately describe the behavior of the variables, thus identifying which are the variables of greatest interest, as well as the points and facts which the customer takes into account when choosing a particular service.

Quality in services is difficult to evaluate, making it imperative to define parameters in order to achieve uniformity in an essentially intangible phenomenon, and this difficulty is felt by RBS-TV, a media company.

In the specific case of the company RBS-TV (Rede Brasil Sul de Televisao) Santa Rosa - RS branch, the main quality instrument is people, i.e. the company's vision of a customer is mainly in the hands of human resources, as they are responsible for the satisfaction or otherwise of the services it provides.

The company that works exclusively in the media has a very direct relationship with its customers when it sells its products, but loses contact with them over the course of the period. Most of the time, the loss of contact with customers without an evaluation of the quality of the services provided to their clientele does not provide an understanding of the service offered in order to check whether it has met their needs and their level of satisfaction.

In order for the company to have an answer to the level of quality and satisfaction of the services it provides, one of the quickest and most widely used methods today is field research, in which, after a careful selection of the method to be used in its execution, the company begins to draw up the structured questions that will provide the answers used in the analysis and subsequent decision-making by the company.

RBS-TV is a company that works exclusively with the media, and the need arose to verify the relationship between sales and after-sales of the services offered in the market, because through this relationship the company will have a view of the degree of customer satisfaction after receiving a service.

It is worth noting that field research is generally not carried out within the broadcaster, and when it is carried out it covers a national or state context. Most of the time, the research carried out by the company is more concerned with the viewers' audience than with the services it provides, which leaves it vulnerable in relation to its competitors by not building customer loyalty.

The RBS-TV branch in the northwest region of Rio Grande do Sul is located in the city of Santa Rosa. It covers 68 municipalities, of which only five have micro-branches, because they are more representative municipalities than the others. As such, the aforementioned locations are a limiting factor in this research.

2. *CLUSTER* ANALYSIS

Conglomerate analysis, also called *cluster* analysis, *is* a technique used to classify objects or cases into relatively homogeneous groups called *conglomerates*. The objects in each conglomerate tend to be similar to each other, but different from objects in other conglomerates. There is no *a priori* information about the composition of the group, or conglomerate, for any of its objects, but it is suggested by the data.

Cluster analysis is widely used in various fields of knowledge, as it is a continuous measure that makes it possible to interpret each individual group and the relationship that this group has with the others.

Agglomeration processes can be hierarchical or non-hierarchical. In hierarchical agglomeration, an order or tree-shaped structure is established, which produces a sequence of partitions into increasingly larger classes. This is not the case with non-hierarchical agglomeration, which directly produces a partition into a fixed number of classes.

Hierarchical processes can be agglomerative or divisive. In agglomerative processes, each object is placed in a separate conglomerate, forming larger and larger groups. The process continues until all the objects are members of a single conglomerate.

Agglomerative methods are commonly used in *marketing* research. They are divided into linkage methods, variance methods and centroid methods. In the *linkage* method, objects are grouped based on the calculation of the distance between them, which can be minimal; the *single linkage* rule, where the distance between the group formed and another is equal to the smallest of the distances between the elements of the two groups, *based on the* maximum *distance*; *Complete Linkage, where* the distance between the group formed and another is equal to the greater of the distances between the elements of the two groups, or based on the average distance between all pairs of objects, where each member of the pair is extracted from each of the conglomerates (*Average Linkage.*

The variance method seeks to generate conglomerates in order to minimize the variance within the conglomerate. *Ward*'s process is a widely used variance method, as it consists of minimizing the square of the Euclidean distance to the cluster averages, i.e. merging the two classes for which the loss of inertia is smallest. The two closest classes must be brought together, taking the loss of inertia incurred by grouping them as the distance between them. In the centroid method, the distance between two clusters is the distance between their centroids (average of all the variables in the group).

The non-hierarchical process (*k-means clustering*) initially determines or assumes a center cluster and then groups together all the objects that are less than a pre-established value from the center. It includes the Sequential Thresholding, Parallel Thresholding and Optimizing Partitioning methods.

However, the most common method is hierarchical classification, where objects are grouped similarly to a taxonomic classification and represented in a graph with a tree structure, called a dendogram. In order to carry out this classification, it is necessary to mathematically define what is characterized as proximity, i.e. the distance between two objects, and then define the criteria for grouping two classes. Among the most common measures used to establish the concept of distance between two objects m and n based on the values of i variables, we can highlight the following forms of measurement:

1ª) *Pearson*'s Linear Correlation Coefficient ;

2ª) Euclidean distance;

3a) Distance from *Manhattan*;

4a) *Mahalanobis* distance;

5a) *Chebychev* distance.

Pearson's correlation coefficient can be measured using the following algebraic expression:

$$r_{nm} = \frac{\sum_{i=1}^{I}(X_{in} - \overline{X}_n)(X_{in} - \overline{X}_m)}{\sqrt{\sum_{i=1}^{I}(X_{in} - \overline{X}_n)^2 . \sum_{i=1}^{I}(X_{im} - \overline{X}_m)^2}} \tag{2.1}$$

On the other hand, the Euclidean distance is the usual measure when using the *cluster* analysis technique. It can be calculated based on the square root of the sum of the squares of the differences in the values of each variable analyzed.

$$D_{n,m} = \sqrt{\sum_{i=1}^{I}(X_{in} - X_{im})^2} \tag{2.2}$$

The *Manhattan* distance can be measured based on the sum of the absolute values of the differences for each of the variables between two objects.

$$D_{nm} = \sum_{i=1}^{I}|X_{in} - X_{im}| \tag{2.3}$$

The *Mahalanobis* distance is measured based on the covariance matrix, as can be seen in the following expression:

$$D_{nm} = \sqrt{(X_n - X_m)C^{-1}(X_n - X_m)'} \tag{2.4}$$

where C is the covariance matrix, which in turn can be expressed as follows:

$$C = |C_{ij}|$$

However, the component of the matrix c_{ij} can still be seen as follows:

$$C_{ij} = \sum_{K=1}^{n} \frac{(X_{ij} - \bar{X})(X_{jk} - \bar{X}_j)}{n} \quad (2.5)$$

The Chebychev distance can be calculated based on the distance between two objects using the absolute value of the greatest difference between the values for any variable analyzed.

As Pereira (2001) points out, *cluster* analysis can be summarized based on the following procedures:

a) Calculation of the Euclidean distances between the objects studied in the multi-plane space of all the variables considered. Therefore, the Euclidean distance can be calculated using the expression:

$D = \sqrt{(x_2 - x_1)^2 + (y_2 - y_1)^2}$, where the pairs (x_1,y_1) and $(x_2, y_{(2)})$) are the coordinates of any points in the plane.

Through these points, we can understand the principle of geometric proximity, the application of which in multivariate analysis is seen as *Cluster* Analysis;

b) Geometric proximity grouping sequence. In this type of analysis, the distance between objects within a space is calculated and grouped according to their proximity, forming a group of the two closest objects. The next step is to check which object is closest to the first group and build a new group, and so on, until all the objects have been brought together in the total group of all the objects studied.

Since objects are linked in order of proximity, i.e. by the shortest distance, it is also necessary to define the notion of distance between groups of objects, or in particular between an object and a group. The most common procedures are: insertion by nearest neighbor, insertion by farthest neighbor and insertion by average distance;

c) Recognition of the clustering steps for the coherent identification of groups within the universe of objects studied;

d) Perform the analysis using a statistical package.

Generally, the variables are grouped according to the correlation between them, and the result obtained is shown by means of a dendrogram, where it is easy to see the *clusters* formed and, from them, draw the conclusions provided by the data collected.

When a *cluster* analysis is used, the distances between the objects studied are shown on the y-axis for all the measurements taken (variables) and the objects that have been grouped are shown on the x-axis.

After the agglomeration scheme, the visual representation of the steps defined at each agglomeration stage is made graphically, where the process resembles a walk, in which the closest objects can be reached based on short steps, while the more distant ones require longer steps, or jumps, to be reached (Pereira, 2001).

However, the literature recommends confirming the results of a classification using factor analysis or correspondence analysis, where both techniques are complementary to cluster analysis, making it possible to interpret the groupings obtained through the classification more quickly, depending on the characters evaluated.

3. FACTOR ANALYSIS

Factor Analysis (FA) began at the beginning of the 20th century with Karl Pearson and Charles Spearman, who studied measures of intelligence. The technique did not spread more quickly due to the difficulty in performing the calculations, which were made easier with the advent of computers. Mathematically, FA is similar to multiple regression analysis in that each variable is expressed as a linear combination of underlying factors (Malhotra, 2001).

The aim of FA is parsimony, as it seeks to define the relationship between variables in a simple way and using a smaller number of factors than the original number of variables.

In short, it is a method essentially used to reduce and summarize data, and it is widely used in *marketing* research:

- Market segmentation to identify the latent variables according to which consumers are grouped;
- Research into a product, with the aim of determining which attributes of a brand influence consumer choice;
- Advertising study, which aims to define the consumption habits of the target market;
- Price study, which can be used to identify the characteristics of price-sensitive consumers.

Therefore, its main purpose is to interpret the structure of a multivariate data set, based on the variance-covariance matrix or the correlation matrix.

When generating the factors, attention must be paid to the measurement scales used for the variables, as there will be occasions when they cannot be directly compared. To do this, you will first need to standardize the variables, so that once they have been transformed they have a zero mean and unit variance, which is achieved by using the "Z" transformation.

In this way, the variance-covariance matrix obtained will be identical to the matrix of correlation coefficients between the variables that have been transformed. However, this standardization has a strong influence on the structure of the variance-covariance matrix and, consequently, on the results of the factor analysis, so its use must be judicious, taking into account the nature of the data and the approach to be taken. Assuming that the variables to be standardized are expressed using the following factor model, these variables can be represented as follows:

$$X_i = A_{ij}F_1 + A_{i2}F_2 + A_{i3}F_3 + \ldots A_{im}F_m + V_iU_i \qquad (3.1)$$

where:

X_i : i-th standardized variable;

A_j: standardized multiple regression coefficient of variable i on common factor j;

F: common factor;

V_i: standardized regression coefficient of variable i on single factor i;

U_i: single factor for variable i; m: number of common factors;

On the other hand, the factors to be extracted can be represented algebraically as follows:

$$F_i = W_{i1} X_1 + W_{i2} X_2 + W_{i3} X_3 + \ldots + W_{ik} X_k \quad (3.2)$$

where,

F : estimate of the ith factor,

W : weight or coefficient of the factor score, k: number of variables.

In fact, FA does not refer to just one statistical technique, but rather to a variety of techniques to make the observed data easier to interpret, i.e. the interrelationships between variables are analyzed so that they can be conveniently described through a group of categories, which we saw earlier as factors.

This shows the importance of the method, which is widely used to facilitate and create variables when necessary, without losing a large amount of information on the primary variables. The method also takes into account the variability expressed by a set of variables, by using a smaller number of index variables or factors. It is assumed that each of the original variables can be expressed as a linear combination of these factors, plus the residual term which represents the dependence of one variable on the others, and these variables can be expressed algebraically as shown below:

$$X_1 = a_{11}F_1 + a_{12}F_2 + \ldots a_{1m}F_m + \varepsilon_1,$$
$$X_2 = a_{21}F_1 + a_{22}F_2 + \ldots a_{2m}F_m + \varepsilon_2,$$
$$\vdots$$
$$X_p = a_{p1}F_1 + a_{p2}F_2 + \ldots a_{pm}F_m + \varepsilon_p, \quad (3.3)$$

where:

a_{ij}: constants,

F_i: common factors or latent variables,

ε : random vector called error or specific factors.

For this purpose, Principal Component Analysis (PCA) is used, although there are several methods that can be used to determine the factors.

When there are a large number of latent variables, there is no need for a direct verification of the factor model on $X_1, X_2..., X_p$. However, making some additional assumptions about the vectors F and ε, this model implies the existence of some relationships involving the covariances, assuming that $E(F) = 0$ and $Cov(F) = E(FF') = I$, so we have that:

$$E(\varepsilon) = 0 \quad e \quad Cov(\varepsilon) = E(\varepsilon\,\varepsilon') = \psi = \begin{bmatrix} \psi_1 & 0 & \cdots & 0 \\ 0 & \psi_2 & \cdots & 0 \\ \vdots & \vdots & \ddots & \vdots \\ 0 & 0 & \cdots & \psi_p \end{bmatrix} \qquad (3.4)$$

Since F and ε are independent, then $Cov(\varepsilon, F) = E(\varepsilon F) = 0$. So the covariance structure for the factorial model can be seen as follows:

$Cov(X) = LL' + \psi$ ou,

$Var(X_i) = \ell_{i1}^2 + \ell_{i2}^2 + ... + \ell_{im}^2 + \psi_i$,

$Cov(X_i, X_k) = \ell_{i1}\ell_{k1} + ... + \ell_{im}\ell_{km}$, (3.5)

$Cov(X,F) = L$ ou $Cov(X_i, F_k) = \ell_{ik}$. (3.6)

According to Malhotra (2001), PA follows a number of steps that can be highlighted as follows:

1°) Problem formulation;

2°) Construction of the correlation matrix;

3°) Determination of eigenvalues and eigenvectors;

4°) Rotation of factors;

5°) Interpretation of the factors;

6°) Calculation of factor scores and selection of substitute variables;

7°) Determining the fit of the model.

The factor model is based on the assumption that variables can be grouped by their correlations, i.e. it is assumed that all the variables within a group are highly correlated, but that they have relatively small correlations in relation to the variables in different groups (Johnson & Wichern, 1998, p.514; Scremin, 2003).

Figure 04 shows the grouping of nine variables ($X_1, X_2,..., X_9$) into three distinct groups (Factor 1, Factor 2 and Factor 3). The variables X_1, X_4, X_5 and X_8, which are highly

Correlated with each other, they form a group and represent an underlying variable, or common factor. Similarly, the variables X_2, X_6 and X_9 define another factor, while the variables X_3 and $X_{(7)}$

form a third factor. In each case, the subset of variables can be interpreted as a manifestation of an abstract underlying dimension - a factor. In this way, the size of the problem is simplified and only three factors are used, which contain most of the information inherent in the original set of nine variables, thus making it easier to interpret the data (Kachigan, 1982, p.237 *apud* Scremin, 2003).

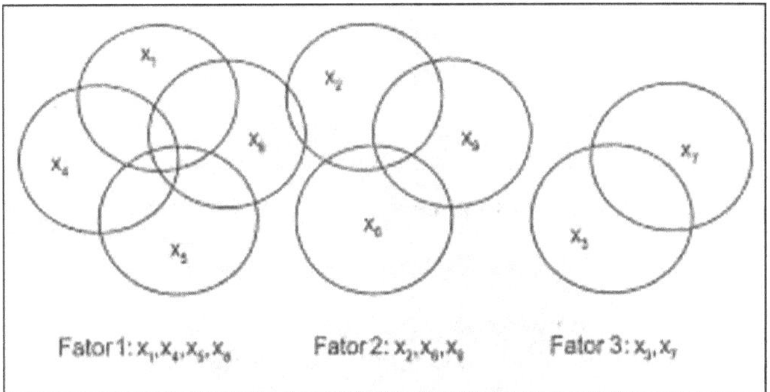

SOURCE: Kachigan (1982), p. 237

FIGURE 04 - Illustration of three underlying factors of a group of nine variables

Figure 04 shows that some factors have variables that are correlated with the variables of another factor, and the formation of these factors will depend on the degree of correlation between the variables. Some concepts considered useful for interpreting PA according to Pereira (1999):

a) *Factor loading:* is the measure of correlation between the derived function and the original measurements. The square of the *factor loading* is the proportion of the variance of the variable that is explained by the factor. It can be interpreted analogously to *Pearson*'s correlation coefficient;

b) The *factor score:* this is the measure taken by the objects studied in the function derived from the analysis. The more the function is derived, it can be understood as the coordinates of each object studied;

c) The *eigenvalue* is the measure of how much of the total variance of the measurements taken can be explained by the factor. It corresponds to the sum of the squares of the *factor loadings* of the functions (factors) derived. In other words, the *eigenvalue* assesses the factor's contribution to the model constructed by factor analysis, with a small value suggesting a small contribution from the factor in explaining the original variables;

d) *Communatity: this is* the measure of how much of the variance of a variable is explained by the factors derived by factor analysis. It corresponds to the sum of the squares of the *factor loadings of* the variable in each of these factors. In other words, *communality* assesses the contribution of the variable to the model constructed by factor analysis, with a low communality suggesting a modest contribution from the variable;

e) The *factor* matrix: this is the correlation matrix between the original variables and the factors that were found, i.e. a matrix describing the *factor loadings* for each original variable. Normally, the matrix used to interpret the results of a factor analysis is a "rotated matrix", which is nothing more than an artifice to promote greater distinction between the relationships found.

According to Macedo (2001), the purpose of FA is to reduce the dimensionality of the system, preserving the initial configuration as best as possible. In other words, it seeks to find the best graphical representations of the multidimensional structure, trying to preserve the original aspect in the best possible way.

One way in which the data set can be better visualized after undergoing factor analysis is by rotating the factors, without the data set losing its total variability. Rotation does not compromise the factors' level of explanation of the variables and facilitates the allocation and construction of each factor, which demonstrates that rotation serves as an analytical device and does not influence communality.

Although reducing dimensionality often brings benefits, it should be borne in mind that the analysis and interpretation of the results can be impaired if the researcher does not have a good knowledge of the subject under study.

Often the interpretation of PA is straightforward, simply by looking at the perceptual maps of the factors. When this interpretation becomes difficult, factor rotation is used, which is nothing more than rotating the axes that represent the factors to be analyzed, so that each factor represents the variables that are not represented in the other factor, and so on. So that the total variability of the system is not altered.

When rotating the factors, it would be interesting for each factor to have non-zero loadings or coefficients for just a few variables, and for each variable to have non-zero or significant loadings with a few factors, if possible just one, so that the variance explained by the individual factors is redistributed by rotation. These include:

- Orthogonal: rotation of factors in which the axes are kept at right angles, thus ensuring no self-correlation between the factors;

- Varimax: orthogonal factor rotation method that minimizes the number of variables with high loadings on a factor, reinforcing the interpretability of the factors. It is most commonly used in applications and evaluates the variances of the loadings within each factor, because as well as keeping the factors orthogonal, it distributes the factor loadings on the axes,

- Oblique: rotation of factors when the axes are not kept at right angles, this type of rotation should be used when you want to measure the degree of correlation of one factor with another, often used in the field of psychology to determine patterns of behavior.

In order to apply factor analysis, it is necessary to test whether the data is linked enough to carry out the analysis, so that FA performs well.

The *Kaiser-Meyer-Olkin Measure of Sampling Adequacy* (KMO) method was used to measure the adequacy of the data.

$$\text{KMO} = \frac{r_1^2 + r_2^2 + r_3^2 + \ldots + r_n^2}{(r_1^2 + r_2^2 + \ldots + r_n^2) + (r_{11}^2 + r_{12}^2 + \ldots + r_{kn}^2)} \tag{3.7}$$

where:

$r_1, r_2, \ldots r_n$: correlation of variables; $r_{11}, r_{12}, \ldots r_{kn}$: partial correlations.

Your critical values are:

- Approximately 0,90: adequacyoptimal;
- Approximately 0,80: adequacygood;
- Approximately 0,70: adequacyreasonable;
- Approximately 0,60: adequacymediocre;
- Approximately 0.50 or less: unsuitable.

The *KMO* is a test that examines the fit of data, taking all the variables simultaneously, and provides synthetic information about the data.

Another test used to verify the assumptions of FA is *the Bartlett Test of Sphericity* (BTS), which tests the hypothesis that the correlation matrix is an identity matrix (diagonal equal to unity and all other measures equal to zero), i.e. that there is no correlation between the variables (Pereira, 1999).

To check the degree to which the items in a questionnaire are interrelated, the *Crombach*'s Alpha statistic is used. *This* is calculated using the variance of individual items and the covariances between items, i.e. it is used to assess the internal consistency of each factor, which is the arithmetic mean of all the half-and-half coefficients resulting from the different ways of dividing the items in the scale in half. This coefficient ranges from 0 to 1, and values below 0.6 indicate unsatisfactory internal consistency.

According to Hair *et al.* (1999), *Crombach*'s alpha (α) has a positive relationship with the number of items in each factor, so alpha values should not be compared if the factors have different numbers of items, as its value tends to increase as the number of items on the scale increases. α can be written as a function of the number of items in the test and the average inter-correlation between the items, and can be calculated as follows:

$$\alpha = \frac{N.\bar{r}}{1+(N-1).\bar{r}} \tag{3.8}$$

Where: N is the number of items, and *r-bar* is the average inter-correlation between the items.

The items in the scale can be divided into two halves randomly, or by a process based on the odd or even items in the set of values. A high correlation between the halves indicates high internal consistency, while this consistency refers to the level of interrelatedness of the scale items.

To determine the factors, the Principal Component Analysis (PCA) technique, which is one of the most widely used methods for finding factors, is used by means of an orthogonal linear transformation of a *p-dimensional* space into a *k-dimensional* space, with $K \leq p$, reducing the number of characters, allowing geometric representations of the individuals and the characters.

Reduction is only possible if the initial "p" characters are not independent and have non-zero correlation coefficients, known as principal component analysis (PCA) Hotelling (1933), Karhunen (1947), Loève (1963), (Johnson & Wichern, 1992).

The main idea is that the first K variables, i.e. the new variables (y) which are now the main components, account for most of the variability in the original data, thus allowing us to stop computing (p-K) components of lesser importance, which results in a reduction in the number of variables, without a considerable loss of information, and being able to express the same set of results. This reduction in the number of characters is not achieved by simply selecting some of them, but by constructing new characters obtained by combining the initial characters using the factors (Bouroche & Saporta, 1980).

PCA is a mathematical technique, basically a linear combination of the original variables, which are written as eigenvalues (λ) and eigenvectors ($\pounds$). The eigenvalues of a correlation matrix represent the variability of each component, and the eigenvectors are the basis for constructing factor loadings. The sum of the eigenvalues equals the number of variables, because the eigenvalues are the variability of each component, and the sum of the components explains 100% of the data, without loss of information, if the components are extracted from the correlation matrix.

If B is a square matrix of dimension (p x p), is it possible to find a scalar (λ) and a vector X of dimension (p x 1) that are not null, by taking the vector X as a common factor on the right, so that the matrix operation is possible, such that:

$$B X = \lambda X,$$

$$B X - \lambda X = 0,$$

$$(B - \lambda I) X = 0 \tag{3.9}$$

To find another equation to complete the system, the condition that the eigenvectors are normalized is determined. This means, in algebraic terms, that the sum of the squares of the elements of the vector must be equal to 1.

If we consider the random vector **X'** = [X_1, X_2,..., X_P], we find the eigenvalues (λ), which are useful for interpreting the constant density of ellipsoids, where we have the covariance matrix Σ, with the following eigenvalues, λ_1 $\lambda_2 \geq \ldots \geq \lambda_{(p)} \geq 0$.

Linear combinations are uncorrelated combinations of Y_1, Y_2,...,Y_P, whose variance is as large as possible. Thus, the first component is the linear combination with the maximum variance, i.e. $Var(Y_1) = \ell_1' \Sigma \ell_1$. To eliminate this indeterminacy, it is convenient to restrict the vector of coefficients to unity. We can then define:

CP_1: Linear combination $\ell_1' X$ that maximizes $Var(\ell_1' X)$ subject to the constraint of $\ell_1' \ell_1 = 1$;

CP_2: Linear combination $\ell_2' X$ that maximizes $Var(\ell_2' X)$ subject to the constraints of $\ell_2' \ell_2 = 1$ $Cov(\ell_2' X, \ell_2' X) = 0$;

CP(i-th): Linear combination $\ell_i' X$ that maximizes $Var(\ell_i' X)$ subject to the constraints of $\ell_i' \ell_i = 1$ and $Cov(\ell_i' X, \ell_k' X) = 0$ for $k < i$.

The eigenvectors of the correlation matrix (**R**) are the essence of the principal component method, as they define the directions of maximum variability and specify the variances. They also serve as weighting factors that define the contribution of each variable to a principal component. An alternative would be to determine the principal components using the variance-covariance matrix (**Σ**), but this could lead to problems with the influence of the sampling units that each variable carries since when using the matrix (**Σ**) there is no standardization of the variables, which does not eliminate the effect of the sampling unit as occurs when using *Pearson*'s correlation matrix (**R**).

In *marketing* research there can be a number of variables, most of which are correlated, which need to be reduced to such a level that conclusions can be drawn quickly, and which denote the reality of the whole set of factors involved in the research.

Each principal component explains a proportion of the total variability, and this proportion can be calculated using the quotient between the value $JI(K)$ and the trace of the variance-covariance matrix (trR) or the correlation matrix. This ratio is called the proportion of total variability explained by the K-th component, and is calculated using the following expression (Pla, 1986):

$$\frac{\lambda(k)}{trR} = \text{explained variation} \qquad (3.10)$$

where R is the correlation matrix that generated the eigenvalues.

The number of components to be used is defined using various criteria, the two most commonly used being Kaiser and Cattel. It is possible to calculate as many principal components as there are variables, but this does not result in any savings. Several processes have been suggested to determine the number of components:

a) Determination based on eigenvalues: only factors with eigenvalues greater than unity are retained; the other factors are not included in the model. An eigenvalue represents the amount of variance associated with the factor. Therefore, only factors with variance greater than unity are included. This criterion was suggested by Kaiser (1960) *apud* Mardia (1979). It tends to include few components when the number of original variables is less than twenty;

b) Determination based on a slope plot: a slope plot is a graphical representation of the eigenvalues *versus the* number of factors in the order of extraction. The shape of the graph is used to determine the number of factors. In general, the number of factors determined by a gradient graph will be one or a few more than that determined by the eigenvalues criterion. This criterion, which takes into account the components prior to the inflection point of the curve, was suggested by Cattel (1966);

c) Determination based on the percentage of variance: the number of factors extracted is determined so that the accumulated percentage of variance extracted by the factors reaches a satisfactory level. Generally, when selecting the principal components, around 80 to 90% of the total variability of the data is retained, which, as Hair Jr. *et al.* (1998, p.104) show, the percentage level to be chosen varies according to each application, with an accumulated variance of around 95% being accepted for applications in the natural sciences, but around 60% or less in the social sciences. Pereira (1999, p.128) recommends using at least 60% when *there is* no standard value for the cumulative percentage of variance explained, and it is up to the researcher to decide;

The interpretation of the selected factors is not always easy to understand and is one of the most delicate points of data analysis, since what you have is a construct that represents a linear combination of the original variables. To this end, the correlations between the original variables and the factors are studied.

Visualizing the variables and the individuals graphically also helps the researcher to better interpret the factors. In this case, the components can be interpreted in two ways: on the one hand, the correlations with the initial characters and, on the other, typical individuals (Bauroche & Saporta, 1980).

4. APPLICATIONS AND DISCUSSIONS

In the commercial context, and especially in the field of service provision, regardless of its characteristics, one of the basic foundations of any activity is customer satisfaction (Azevedo & Takaki, 2003).

Therefore, improving the services a company provides to its customers is a fundamental step towards business success. With accurate information about their customers' perceptions of service quality, organizations can make better decisions to serve them better (Hayes, 2001).

In order to understand the behavior of the variables under study, an analysis of the questionnaires applied to customers was carried out. In the first part, a profile of the advertising customer was drawn up, as well as a descriptive study, and in the second part, multivariate analysis techniques were used. The company for this study is RBS-TV, which will be able to analyze the quality of its services, the satisfaction of its customers and the development of plans for after-sales service.

The population under study is made up of 634 companies, of which 155 make up the sample evaluated, taking care to size the sample using an error of 5%, and values of "p" and "q" equal to 50%.

The questionnaires were administered in the municipalities of Santa Rosa, Très Passos, Três de Maio, Horizontina, Santo Ângelo and Sao Luiz Gonzaga. The city of Santa Rosa is home to the branch office for the northwest region of the state of Rio Grande do Sul, and the other cities have RBS-TV micro-branches, and the periods of data collection are not highlighted to avoid possible identification of the individuals in the survey.

4.1 A view of the company - first part of the questionnaire

Of RBS-TV's advertisers, 61% are small and medium-sized companies in the commercial sector, with the most frequent number of employees being five. These companies have been operating in the market for more than 10 years, which means they are solid companies, and almost 100% of them are privately owned.

The media that the interviewees use the most is radio, and 40% of the companies use more than one media. When asked which media brings the most return, 70% of the companies interviewed said television, i.e. they advertise more on the radio because it is more affordable.

The frequency of advertising by the companies is without a set date; they advertise according to the company's needs and the time of year when they need to advertise their services and products more. As a result, 54% of the companies surveyed were not being advertised at the time of the survey, and an open question was used to find out why they were not being advertised. Among the most frequent answers were: the high price and the company's lack of resources to invest in media.

With regard to annual media investment, 42.6% of the companies invest between R$1,000.00 and R$5,000.00, i.e. an investment considered to be low, which can be justified by the fact that the advertising companies are mostly small and medium-sized.

In order to carry out a more in-depth analysis of RBS-TV's customer profile, we cross-referenced the variables shown in the following tables. Table 01 shows the cross-referencing between the company's branches of economic activity and the frequency with which it advertises on RBS-TV.

TABLE 01- Cross-referencing of the company's branch of economic activity and the frequency of the ad.

Ad frequency					
Activity	**Monthly**	**Half-yearly**	**Annual**	**No date**	**Total**
Commercial	30	14	7	44	**95**
Industrial	2	4	1	5	**12**
Provider	7	3	3	20	**33**
Education	0	0	1	7	**8**
Others	1	0	1	5	7
Total	**40**	**21**	**13**	**81**	**155**

The table above shows that 46.3% of commercial companies advertise on RBS TV without an established date, and this percentage rises to 60.6% for service providers.

Table 02 shows the cross-referencing of the variable by branch of economic activity of the company analyzed versus the annual investment made by these companies.

TABLE 02 - Cross-referencing the type of company and annual investment (in thousands).

		Annual Investment					
Activity	**Up to 1**	**1 à 5**	**5 à 10**	**10 à 50**	**More than 50**	**Others**	**Total**
Commercial	16	36	14	20	8	1	**95**
Industrial	1	6	2	2	0	1	**12**
Provider	6	17	7	3	0	0	**33**
Education	0	4	3	0	1	0	**8**
Other	1	3	2	1	0	0	7
Total	**24**	**66**	**28**	**26**	**9**	**2**	**155**

Table 02 shows that 38% of the companies in the commercial sector and 52% of the service providers invest between R$1,000.00 and R$5,000.00 a year, thus demonstrating that the investment is considerably low, given the economic reality of the region and the fact that these are small and medium-sized companies, as mentioned above. It can also be seen that 21% of commercial companies advertise between R$10,000.00 and R$50,000.00 annually, a significant percentage compared to the reality of other companies.

On the other hand, Table 03 shows the cross-referencing of the variable classifying the company according to its size and investments. It can be inferred that 56% of small companies advertise on TBS-TV and spend between R$1,000.00 and R$5,000.00 annually, which is not the case with medium-sized companies, where the percentages are practically equal in terms of investments, i.e. 30% invest between R$1,000.00 and R$5,000.00 annually and 28% invest between R$5,000.00 and R$10,000.00 and R$10,000.00 to R$50,000.00 annually.

TABLE 03 - Cross-referencing the variables classifying companies according to their size and level of investment (in thousands).

Class	**Up to 1**	**1 à 5**	**5 à 10**	**10 à 50**	**More than 50**	**Others**	**Total**
Micro	10	18	4	0	0	1	**33**
Small	11	29	7	4	1	0	**52**
Average	4	16	15	15	3	0	**53**
Large	0	3	2	6	5	1	**17**
Total	**25**	**66**	**28**	**25**	**9**	**2**	**155**

Through the individual analysis of each question, as well as the cross-references between the most significant variables, it can be concluded that the clients who use RBS-TV are private, small and medium-sized companies in the commercial sector, with more than 10 years' experience in the market. They use more than one medium to publicize their company, and they get the best returns from television, but they don't invest more in this vehicle because the cost is too high.

At the time of the survey, just over half of the companies surveyed were not advertised, which leads us to infer that they advertise without an established date, with an annual investment of around R$5,000.00.

Through this profile, RBS-TV will be able to see which customers need to be more concerned, and thus set targets for a substantial increase in the annual investments made by these companies.

4.2 The customers' view - second part of the questionnaire

Below are the results of the second part of the questionnaire, which mainly involves items related to satisfaction with the services provided and after-sales service.

Before looking at the results, below is the questionnaire used, where each question is summarized in one word, for a better interpretation of the variables under study.

Question 1: RBSTV Santa Rosa as a media option - "option";

Question 2: On the return you get from investing in advertising on RBS-TV Santa Rosa - "return";

Question 3: Regarding the alternatives for ads - "ad";

Question 4: The ad proposed by the agency met your needs - "need";

Question 5: The media slots offered catered for your target audience - "slot";

Question 6: Regarding the service provided by the agent - "service";

Question 7: When you contact the company, you receive a response to all your requests, complaints and/or suggestions - "requests";

Question 8: When I book a meeting time, the agent is available for the meeting at a time that suits me - "availability";

Question 9: The agent's promptness when arriving at the meeting - "promptness";

Question 10: Punctuality of the meeting start time - "punctuality";

Question 11: Regarding after-sales service - "after-sales";

Question 12: RBS TV Santa Rosa price list - "price";

Question 13 Condition of payment - "condition";

Question 14: Satisfaction with continuing or returning to advertise on RBS TV Santa Rosa - "continue".

We first determined the descriptive statistics of the variables studied, before applying the multivariate analysis technique, as shown in Table 04.

TABLE 04 - Descriptive statistics of the variables analyzed using a five-point Likert scale.

Variables	N	Average	Deviation Standard	Minimum value	Maximum value
Option	155	4,000000	0,693195	2	5
Return	155	3,696774	0,824733	1	5
Advertisement	155	3,774194	0,810230	1	5
Need	155	3.812903	0,745437	2	5
Schedule	155	3,890323	0,743524	1	5
Service	155	4,425806	0,654142	2	5
Request	155	4,051613	0,700589	1	5
Availability	155	4,167742	0,611782	1	5
Promptness	155	4,238710	0,510715	3	5
Punctuality	155	4,193548	0,645632	1	5
After-sales	155	3,832258	0,903236	1	5
Price	155	2,980645	1,053571	1	5
Condition	155	3,722581	0,793934	1	5
Continuc	155	3,961290	0,710623	1	5

Table 04 shows that the averages of the variables analyzed are around four, i.e. there is a predominance of satisfactory levels, with only the price variable remaining at an unsatisfactory level.

Pearson's coefficient of variation shows that the average of these variables is statistically significant, at around 22%, with the exception of the price variable, where the coefficient of variation is around 33%, showing that price is the variable that reveals the greatest dispersion among the respondents' opinions. The average response for the price variable was 2.98, which shows that the values represented by the very dissatisfied and dissatisfied options influenced the average downwards.

On the other hand, the standard deviation of the variables is considered low, and there is not a great deal of variation between the answers obtained.

The correlation matrix is used to check the correlation between the variables, and factor analysis will use it to derive the factors.

TABLE 05 - Correlation matrix between variables

	Options	Retor	Anùnc	Necessary	Time	Service	Request	Available	Prest	Pontu	Post-Ven	Price	Condi	Cont
Options	1,00													
Retor	**0,51**	1,00												
Anùnc	0,17	0,25	1,00											
Necessary	0,25	0,41	**0,54**	1,00										
Time	0,35	0,40	0,16	0,31	1,00									
Service	0,39	0,24	0,31	0,20	0,28	1,00								
Request	0,29	0,20	0,11	0,09	0,21	0,36	1,00							
Available	0,18	0,18	0,16	0,08	0,28	0,49	0,39	1,00						
Prest	0,24	0,16	0,19	0,15	0,29	0,43	0,44	**0,66**	1,00					
Pontu	0,26	0,12	0,25	0,13	0,23	**0,56**	0,41	**0,71**	**0,75**	1,00				
Post-Ven	0,36	0,28	0,23	0,22	0,24	0,30	0,27	0,22	0,17	0,33	1,00			
Price	0,28	0,32	0,35	0,31	0,41	0,41	0,31	0,18	0,20	0,20	0,30	1,00		
Condi	0,27	0,30	0,35	0,25	0,27	0,30	0,27	0,12	0,13	0,14	0,22	**0,56**	1,00	
Cont	**0,51**	**0,52**	0,35	0,37	0,43	0,40	0,41	0,24	0,33	0,27	0,45	**0,55**	0,46	1,00

Table 05 shows that there is a correlation between the variables, although in some cases this correlation is considered low, which could apparently invalidate the factor analysis.

As a preliminary analysis, the FA will use the cluster analysis procedure, as this will make it possible to identify which variables belong to the same group, i.e. to identify which variables the customer identifies as having the same characteristics for him, thus corroborating with the company in possible formulation of sales strategies for its services.

It should be noted that the objects are grouped based on the calculation of the distance between them. Therefore, Euclidean distance was used to construct the dendrogram, and the linkage method was complete linkage.

Figure 07 shows the dendrogram with all the variables, in which we can identify the formation of three clusters, obtained through a cross-section made at the greatest distance between the clusters, or at the researcher's discretion. The first cluster is formed by the price variable, the second by the variables solic, pontu, prest, dispon e atend, and the last cluster is formed by the variables pos- ven, condi, necess, anunc, hora, retor, cont e opç.

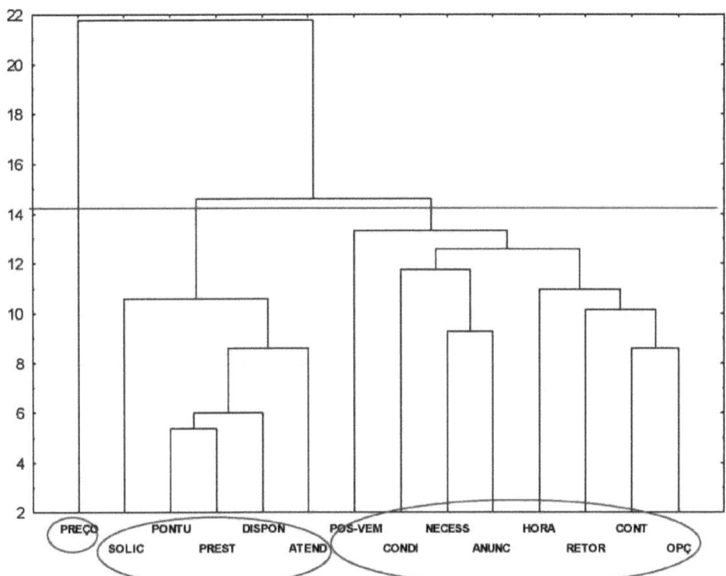

FIGURE 07 - Dendrogram involving all the variables in part 2[(a)] of the questionnaire.

By analyzing the dendrogram, it was decided to remove some variables that have the same representation within the cluster. As only the price variable appears in the first cluster, no changes were made. The second cluster is related to the agent's service, so the variable pont was removed, and in the third cluster the variables anunc and opç were removed, resulting in a new dendrogram, in which three new clusters can be seen, as shown in Figure 08.

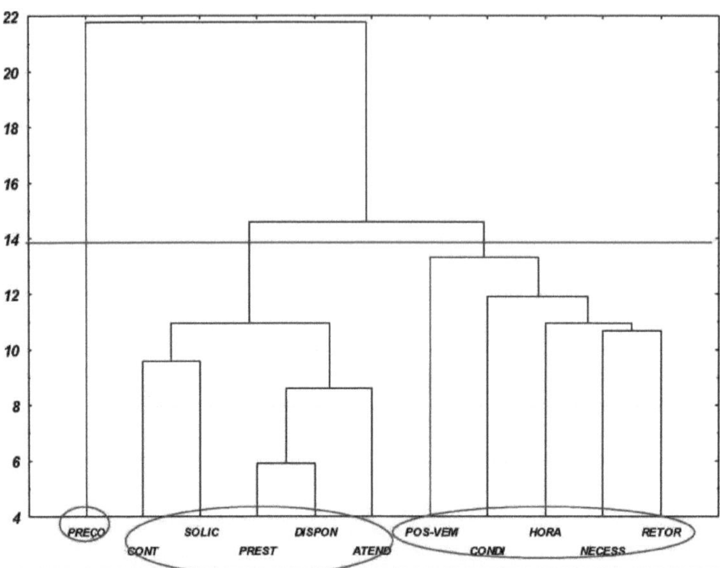

FIGURE 08 - Dendrogram after removing variables with the same degree of relationship.

With the price variable remaining in the first cluster, the variables available and retainer were removed from the second and third clusters respectively, forming a new dendrogram, showing the formation of three new clusters, as shown in Figure 09.

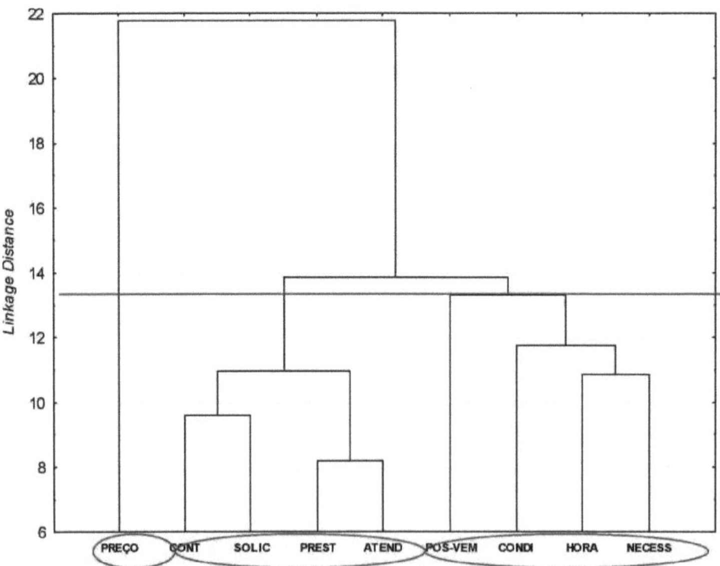

FIGURE 09 - Dendrogram after removing variables with the same degree of relationship.

Since there were still variables with the same profile in the second and third clusters, the variables atend and hour were removed again, forming a new dendrogram, which is shown in Figure 10.

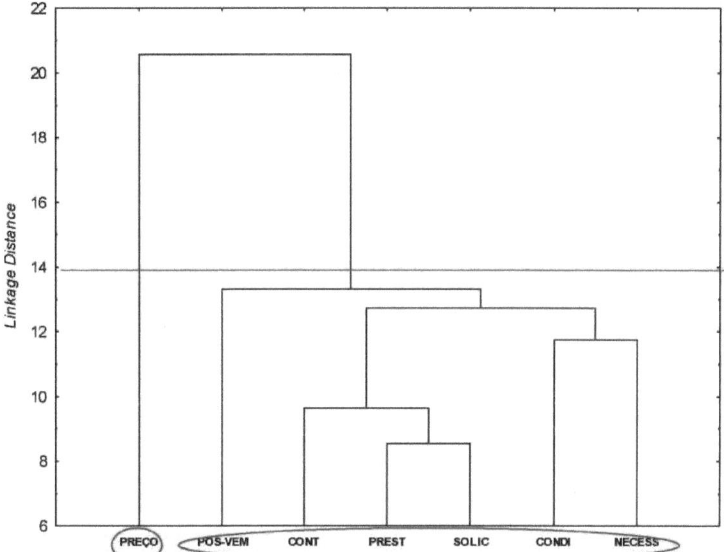

FIGURE 10 - Dendrogram after removing variables with the same degree of relationship.

The dendrogram shows the formation of two clusters, which contain the most relevant variables within the original set of variables surveyed.

Once the variables had been identified as belonging to the same cluster and the variables considered most relevant from the point of view of the company and the researcher remained in the analysis, a factor analysis of these variables was carried out in order to compare how the data set behaves when the entire data set is used, as well as the data set that had been reduced using the cluster analysis technique.

If the result of the reduced data set is satisfactory, the company will be able to use these seven variables for subsequent surveys, which can even be carried out by telephone, thereby reducing the time taken to carry out the surveys, reducing costs and also serving to monitor the company's after-sales services.

After the cluster analysis, the original data is presented, which will be analyzed using factor analysis.

4.3 Factor analysis of the complete data set

In order to study PA, it is essential to calculate the correlation matrix, as was shown in section 4.2, because this way it is possible to know the interrelationship between the variables, as well as showing that a univariate analysis would not be enough to reveal the behavior of the entire set of data under analysis.

For this reason, the KMO was calculated and a value of 0.843 was obtained, suggesting a good fit between the data used, signaling that the use of the factor analysis technique could be carried out. This result is corroborated by Bartlett's test, which gave a value of 860.836, with 91 degrees of freedom and a significance level of $p \ll 0.05$.

The internal consistency of each factor was checked using Crombach's alpha, which gave a value of 0.8564, identifying satisfactory consistency between the factors.

The first step in performing FA is to determine the eigenvalues and the percentage of explanation of each eigenvalue, and then extract the eigenvalues that will represent the set of variables. To do this, the correlation matrix was used, as shown in Table 05.

Table 06 below shows the eigenvalues and the percentage of variance explained.

TABLE 06 - Eigenvalues and percentage of variance explained.

Factors	**Eigenvalues**	**Variance explained (%)**	**Cumulative eigenvalues**	**Var. Explained (%)**
1	**5,080402**	36,28859	5,08040	36,28859
2	**2,014212**	14,38723	7,09461	50,6758
3	**1,124493**	8,03209	8,21911	58,7079
4	**1,012130**	7,22950	9,23124	65,9374
5	0,842840	6,02028	10,07408	71,9577
6	0,689221	4,92301	10,76330	76,8807
7	0,654506	4,67505	11,41780	81,5557
8	0,518466	3,70333	11,93627	85,2591
9	0,454574	3,24696	12,39084	88,5060
10	0,419413	2,99581	12,81026	91,5018
11	0,377150	2,69393	13,18741	94,1958
12	0,333499	2,38214	13.,2091	96,5779
13	0,286643	2,04745	13,80755	98,6254
14	0,192451	1,37465	14,00000	100,0000

According to equation 2.17, we know that the percentage of variance explained by the first eigenvalue is 36.28859%, which represents the total variability of the system, where the main diagonal is made up of values equal to 1.

After extracting the eigenvalues and the percentage of variance explained, it is necessary to decide on the number of factors to be removed for analysis. Therefore, Table 06 shows that 65.9374% of the data is explained by eigenvalues greater than 1, which can be corroborated using the graphical method suggested by Cattel (1966), shown in Figure 11, i.e. the decision was made to use four factors. Therefore, the original 14 variables are replaced by just four factors. It should be noted that there is a loss of information from 100% to 65.94%, but this is compensated for by the reduction in the number of variables to be analyzed, which are now just four factors.

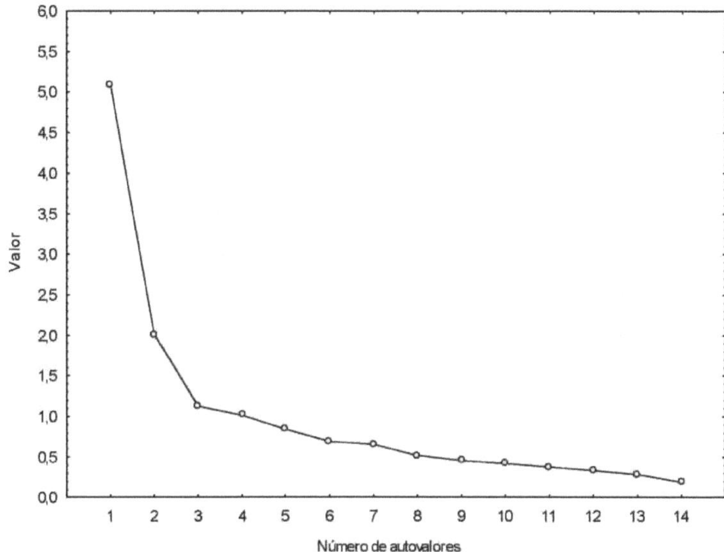

FIGURE 11 - Slope graph of the eigenvalues.

Using the eigenvalues shown in Table 06, the eigenvectors were estimated in order to write the linear combination that will give rise to the factors as shown in Table 07.

In the complete set, there are 14 variables, which correspond to the eigenvalues and eigenvectors described in Tables 07, 08 and 09 respectively.

TABLE 07 - Eigenvectors for writing the linear combination, which will give rise to the factor loadings.

	Factor 1	Factor 2	Factor 3	Factor 4
Options	-0,119870	-0,093379	-0,388532	-0,244524
Retor	-0,113472	-0,194454	-0,224087	-0,364380
Anùnc	-0,099364	-0,134937	0,598151	-0,077390
Necessary	-0,096795	-0,208708	0,400852	-0,383748
Time	-0,114417	-0,078452	-0,172261	-0,163626
Service	-0,135602	0,114850	0,065374	0,114712
Request	-0,112736	0,129036	-0,203456	0,299519
Available	-0,116045	0,306408	0,063972	-0,085977
Prest	-0,122759	0,291155	0,087553	-0,094699
Pontu	-0,127572	0,308926	0,122646	-0,079170
Post-Ven	-0,106728	-0,045173	-0,169256	-0,065900
Price	-0,126552	-0,159309	0,058963	0,446007
Condi	-0,107704	-0,175056	0,086411	0,531238
Cont	-0,149472	-0,141494	-0,167086	0,066691

In order to check the importance of each variable in the composition of the factor, the factor loadings matrix was calculated. Table 08 shows this matrix using the data without rotation.

TABLE 08 - Factor loadings in the factor compositions.

	Factor 1	Factor 2	Factor 3	Factor 4
Options	-0,608987	-0,188086	-0,436901	-0,247490
Retor	-0,576484	-0,391671	-0,251984	-0,368800
Anùnc	-0,504807	-0,271792	0,672617	-0,078328
Necessary	-0,491759	-0,420382	0,450756	-0,388403
Time	-0,581283	-0,158019	-0,193706	-0,165611
Service	-0,688913	0,231332	0,073512	0,116103
Request	-0,572745	0,259905	-0,228785	0,303152
Available	-0,589553	0,617170	0,071936	-0,087020
Prest	-0,623667	0,586449	0,098453	-0,095848
Pontu	-0,648115	0,622243	0,137915	-0,080130
Post-Come	-0,542222	-0,090988	-0,190327	-0,066699
Price	-0,642934	-0,320882	0,066304	0,451417
Condi	-0,547179	-0,352600	0,097169	0,537682
Cont	**-0,759380**	-0,284999	-0,187887	0,067500

From the correlation coefficients identified in Table 08 of the factor matrix, before the rotation, it can be seen that the values are very close and therefore a rotation is necessary so that the factors can be better interpreted, so it was decided to carry out a normalized Varimax rotation to determine the new factors.

Once the eigenvalues and eigenvectors had been extracted, as shown in Tables 06 and 07, the number of factors to be used in the analysis was selected, and it was found that four factors would represent the data set using the selection methods mentioned above. After this stage, the *factor loadings* matrix was examined to see the contribution of each variable to the composition of the linear combination. Table 08 shows that only the cont variable has a significant load of -0.759380.

4.4 Factor analysis of the complete data set with *varimax* rotation

In order to identify the variables with the greatest contribution, the analysis was carried out using a normalized varimax rotation, allowing for a better distribution of the variables, as shown in Table 09.

TABLE 09 - Factor loadings in the factor composition after Varimax rotation.

	Factor 1	Factor 2	Factor 3	Factor 4
Options	0,221531	0,076034	0,076784	0,088129
Retor	**0,908799**	0,027101	0,176180	0,101774
Anùnc	0,069512	0,068193	0,257641	0,139739
Necessary	0,169031	0,043602	**0,921555**	0,075349
Time	0,151071	0,104508	0,120350	0,085737
Service	0,059963	0,174678	0,054225	0,107361
Request	0,051392	0,181868	0,009308	0,101522
Available	0,061755	0,338567	0,002407	0,023228
Prest	0,032148	**0,899164**	0,048690	0,021091
Pontu	-0,003214	0,500990	0,024570	0,027252
Post-Ven	0,087342	0,044603	0,070342	0,059976
Price	0,098312	0,047109	0,104243	0,268289
Condi	0,096515	0,020086	0,074624	**0,927210**
Cont	0,227100	0,124524	0,131118	0,196862

Table 09 shows the four main components: profitability, promptness, necessity and conditions, on which the factor planes will be drawn for a better interpretation. It is worth noting that the variability of the system is not altered when a rotation of this type is carried out, only the coordinates of the axes are rotated and, in this way, the inertia of the system remains unchanged. Therefore, the analyses carried out previously are still valid, but now with a new association between the original variables and the factors.

Once the number of factors to be worked on has been selected, it is then possible to graph these factors in Figures 12, 13 and 14.

Next, the factor planes between the factors are plotted. Figure 12 shows factor 1 versus factor 2.

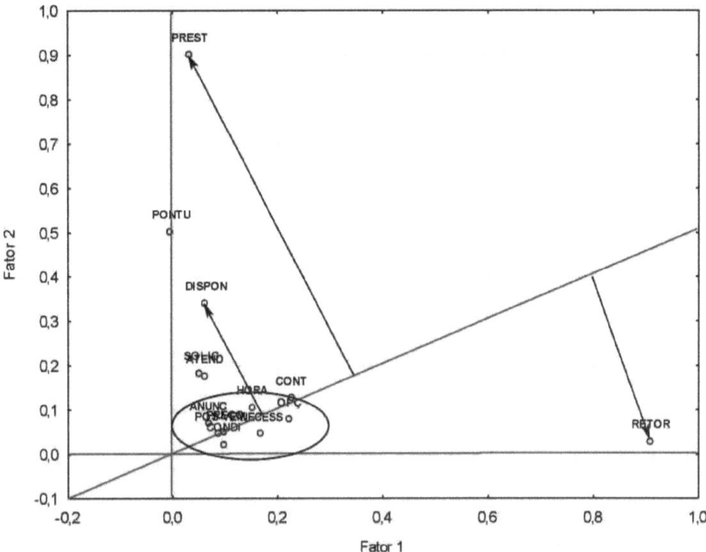

FIGURE 12 - Representation of factor 1 versus factor 2.

The factor pianos show the behavior of the most representative variables. The abscissa axis shows the variable retor, which was asked about the return achieved by investing in advertising on RBS-TV, with an average response of 3.696774 and a factor loading value of 0.908799. The ordinate axis is represented by the variable prest which was asked about the agent's promptness when arriving at the meeting, with an average response of 4.238710 and a loading factor of 0.899164. These variables were the most evident in the analysis.

The variables inside the ellipse are those that have little expression in the composition of the factor, i.e. they are not significant at the 7% level. Therefore, when clients advertise their company on RBS-TV, they take into account the return they will get from the media (extreme point on the x-axis), as well as the agent's promptness (extreme point on the y-axis) when negotiating.

Once the most important variables on each axis or factor have been identified, Factor 1 can be called the financial return to the advertiser and Factor 2 can be called the salesperson's ability to meet the customer's needs, here called promptness.

The main idea of factor analysis is, based on the factor *loadings*, to create labels for the factors according to the most representative variable in the selected factor.

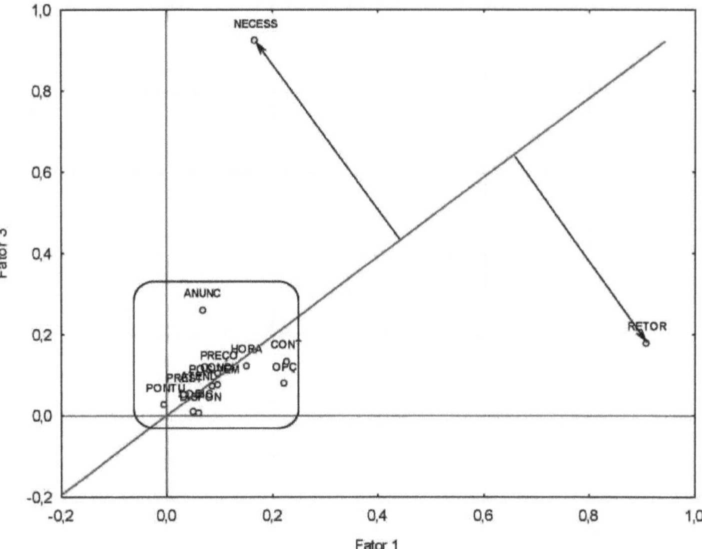

FIGURE 13 - Representation of factor 1 versus factor 3

In the factor plan shown in Figure 13, it can be seen that the variable retor remains on the abscissa, which is the most representative variable, and on the ordinate axis it is the necessary one, with an average response of 3.812903 and *factor loadings* equal to 0.921555. The other variables are quite close to the origin and are therefore not significant. In Figure 3, Factor 3 represents the customer's need to advertise on RBS-TV.

Looking at Figures 13 and 14, we can see that the return variable is the most representative, i.e. the client, when running an ad on RBS-TV, mainly takes into account the return he will get from the day.

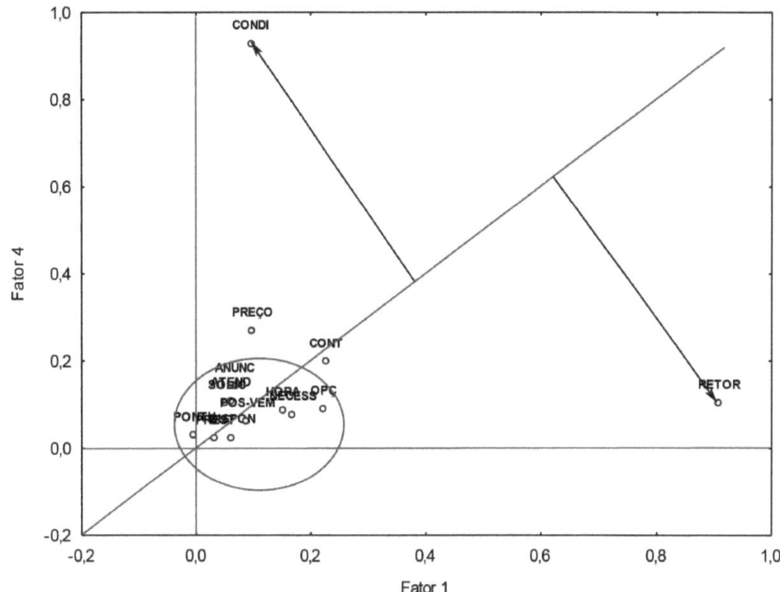

FIGURE 14 - Representation of factor 1 versus factor 4

In this factor plan, the variable condi represents the question about the payment conditions offered by the company, being the most representative on the ordinate axis, and having an average response of 3.722581 and *factor loadings of* 0.927210.

Factor 4 represents the customer's payment terms for the ad, which can be paid in cash or in installments.

After carrying out the factor analysis with the original variables, the FA was carried out with only the reduced variables, using cluster analysis.

First, the eigenvalues and the percentage of explanation of each eigenvalue were determined, followed by the extraction of the eigenvalues that will represent the new set of variables, as shown in Table 10.

TABLE 10 - Eigenvalues and percentage of variance explained.

Factors	**Eigenvalues**	**Variance Explained (%)**	**Cumulative eigenvalues**	**Cumulative explained variance (%)**
1	**2,916719**	41,66741	2,916719	41,6674
2	**1,101831**	15,74044	4,018549	57,4078
3	**0,848400**	12,12000	4,866949	69,5278
4	0,786174	11,23106	5,653123	80,7589
5	0,511933	7,31333	6,165056	88,0722
6	0,433623	6,19461	6,598679	94,2668

7	0,401321	5,73316	7,000000	100,0000

Table 10 shows that the number of factors to be removed is three, which explain 69.5278% of the data, which can be corroborated by the graphical method shown in Figure 15.

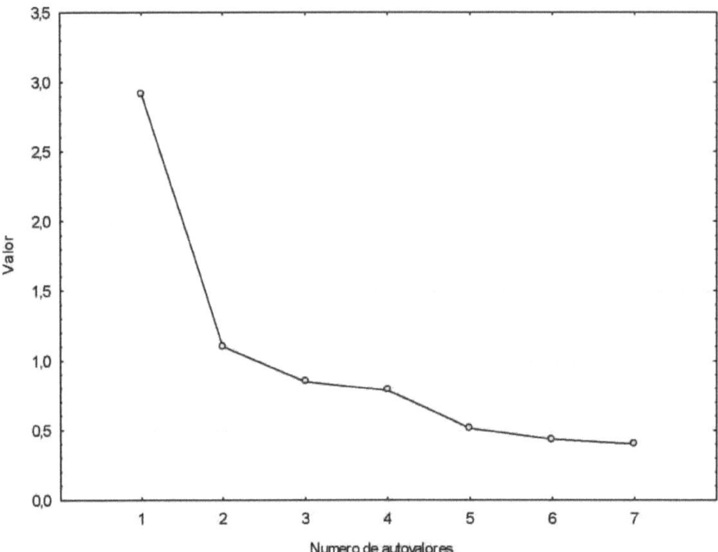

FIGURE 15 - Slope graph of the eigenvalues of the variables selected by cluster analysis.

Another pertinent analysis was the comparison between Tables 06 and 10, after determining the varimax rotation, which reduced the number of components to be selected and increased the degree of explanation by around 6%.

It was also found, after this rotation, that the variables with the greatest contribution to the system are shown in Table 11.

TABLE 11 - Factor loadings in the factor compositions after Varimax rotation.

	Factor 1	**Factor 2**	**Factor 3**	**Factor 4**	**Factor 5**	**Factor 6**	**Factor 7**
Necessary	0,095150	0,060151	**0,973841**	0,090222	0,014172	0,114645	0,132201
Request	0,106029	0,217154	0,014611	0,114022	0,945780	0,111172	0,146728
Prest	0,032905	**0,965856**	0,061839	0,059851	0,203350	0,062523	0,115641
Post-Ven	0,069721	0,060089	0,092192	0,965279	0,107760	0,107316	0,167738
Price	0,275369	0,072827	0,137200	0,125314	0,122330	0,907557	0,213813
Condi	**0,937826**	0,034622	0,103808	0,073834	0,107085	0,249193	0,171730
Cont	0,208523	0,151217	0,176551	0,221196	0,180689	0,238003	0,874220

Table 11 shows a detailed analysis of three main components: *cond, prest* and *necess,* in which factor plans were drawn to visualize and compare the results with the original group of variables.

In the factor plan shown in Figure 16, it can be seen that the variables *prest on* the ordinate axis and *condi* on the abscissa axis are the most representative, as they are furthest from the origin, while the other variables are not very significant.

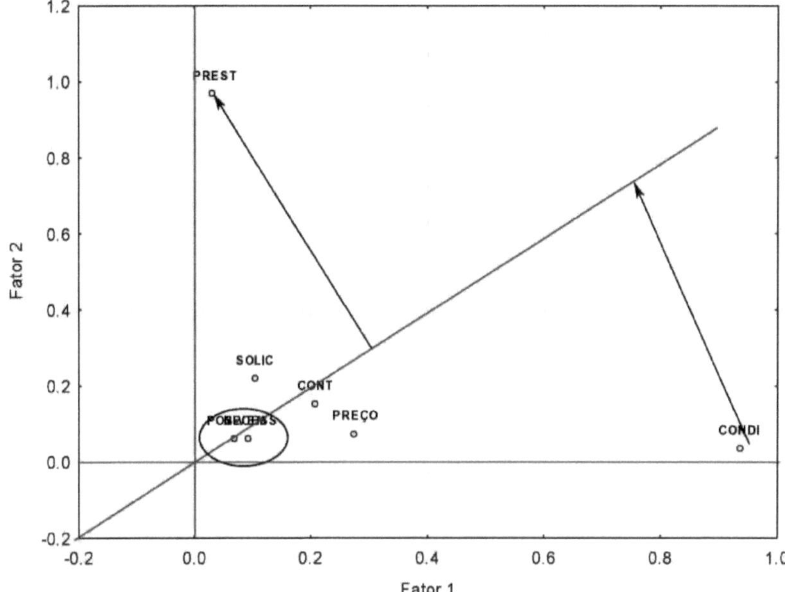

FIGURE 16 - Representation of factor 1 *versus* factor 2.

On the other hand, if you look at the factor plan for factor 1 and factor 3, shown in Figure 17, you can see that the most representative variables are *ness* on the ordinate axis and *condi* on the abscissa axis.

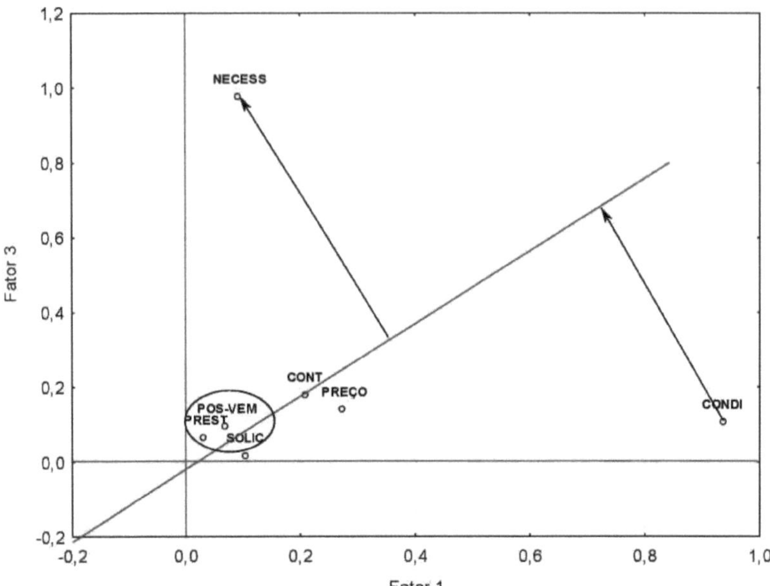

FIGURE 17 - Representation of factor 1 *versus* factor 3

By comparing the factor plans of the complete data set and the variables that were reduced by *cluster* analysis, it can be concluded that the remaining seven variables are representative of the original set and can be used in future research, as they show the same results as those obtained without the reduction.

When analyzing the complete data set, four main components were selected: return on media, agent promptness, whether the ad met the company's needs and payment conditions. As for the set of variables that were reduced, it was decided to use three main components: payment conditions, agent promptness and whether the ad met the needs of the clients. In other words, only the return variable was left out, which has the same meaning as the need variable, because the customer will only be satisfied with an advertisement if they get a return from it.

After analyzing the objective questions, and then applying the techniques mentioned above, we proceeded to check question number fifteen, which is an open question, where the interviewees were asked to make a suggestion about the services provided by the sales department, and they made suggestions related to RBS-TV as a whole. In this sense, the most frequently cited issues can be highlighted:

a) Very weak signal in the city of Sao Luiz Gonzaga;

b) Put in an antenna system so that the inner cities can better receive the signal broadcast by RBS-

TV;

c) Greater diversity of proposals for micro and small companies to advertise more frequently;

d) RBS-TV could broadcast its signal via satellite dishes, given the large number of connected television sets;

e) More attention and responsibility from the collections department;

f) Work with agencies that offer support to the company in putting together the commercial, or RBS-TV has its own agency;

g) The cost of media is too high for micro and small businesses;

h) Review prices according to the local reality of each region;

i) Propose more affordable prices in prime time;

j) Greater flexibility in negotiations depending on the type of client;

k) Putting together a larger number of promotional packages at different times;

l) Visits by the executive department to clients who are highly representative of RBS-TV;

m)More choice of agencies to improve the quality of commercials without increasing media costs;

n) More creativity on the part of agencies when putting together ads;

o) Promotion of partnership campaigns between companies and RBS-TV, allowing companies in general to air their ads at a more affordable price.

5 CONSIDERATIONS

The main considerations following the application of the suggested statistical techniques, as well as pointing out some suggestions for the company, outside the study with regard to improving the quality of the services offered, leading to an increase in the quality of services.

The purpose of this study was to determine which quality factors, according to customers' perceptions, are considered most important when they choose RBS-TV Santa Rosa branch to broadcast their companies' advertisements.

Multivariate statistical methods were used to better identify the most relevant variables.

According to the questionnaire that was applied, it was possible to get a profile of the companies that advertise on RBS-TV, which leads us to say that they are commercial, private, small and medium-sized companies that have been operating in the market for more than 10 years.

The media they use the most is radio, and TV is the one that brings in the most returns for their companies, but they no longer use it because its costs are higher than radio. They advertise without a set date and with an annual investment of around R$1,000.00 to R$5,000.00.

With the data obtained in the second part of the questionnaire, we first used descriptive statistics, which provided an average level of customer satisfaction. Most of the questions were satisfied (4 on the *Likert* scale), with the exception of the price variable, which was dissatisfied (2 on the *Likert* scale).

Multivariate statistical methods were used to extract the principal components, which reduced the number of original variables.

AF was applied after verifying the correlation between the variables and calculating the *KMO*, which provided an adequacy value of around 80%, with a satisfactory internal coherence of 85%, provided by *Crombach*'s Alpha.

Prior to the AF technique, a *cluster* analysis was carried out to identify the variables that belong to the same *cluster*, thus making it possible to see which variables customers identify as having the same effect.

The final analysis, after removing the variables with the same meaning within each cluster, resulted in the formation of two clusters, the first of which contains the *price* variable alone, and the second the variables *"put-sell", "cont", "prest", "solic", "cond"* and *"necess"*.

Subsequently, a factor analysis was carried out on the remaining seven variables, which were verified with the results obtained when a factor analysis was carried out on all the variables, reaching the conclusion that these seven variables could be representative of the original set of variables, which was confirmed by the factor pianos in both cases.

Therefore, in a future survey, RBS-TV could use these seven questions to periodically evaluate the quality of its services.

With the variables selected, using the PCA method it was possible to plot the factorial pianos and reach the following conclusion: RBS-TV customers choose this media outlet in the following order of importance:

1ª) media feedback;

2ª) the agent's promptness;

3a) necessity;

4a) payment terms.

As a suggestion to the company, it is hoped that there will be more direct contact with the statistical procedures that are part of PA, as well as studying the correlation between the variables, so that a better understanding can be obtained of the criteria that customers use to choose the type of media and the time of day that brings the most returns to their companies.

Another aspect of great importance for RBS-TV is after-sales monitoring, as it is through this that it will be able to verify the progress of the services provided and the level of satisfaction with the services received.

From the AF, it can be seen that the satisfaction of the service received by customers is mainly in the hands of the agents, so they are responsible for the good image of the company, and therefore need continuous training in sales and customer service techniques.

BIBLIOGRAPHY

BACHMANN, G.M. **The use of Factor Analysis to determine perceived quality in a university library.** Curitiba: 2002 119 p. Dissertation (Master of Science) - Federal University of Paranà, 2002.

BOUROCHE, J-M.; , SAPORTA, G. **Data Analysis.** Rio de Janeiro: Zahar Editores, 1980.

BERRY, L. **Maximum Satisfaction Services** - A Practical Guide to Action. Rio de Janeiro: Campus, 1996.

CATTEL, R.B. The scree test for number of factors. Multivariate Behavioral Research, 1, 1966, p. 245-276.

DAFFY, C. O. Six Sigma in customer service. **Banas Qualidade, Sao Paulo, n. 142, p.** 30-36, March 2004.

FRIEDRICH, A.; QUADROS, F.S.; VIEGAS, N. Measuring consumer satisfaction in Porto Alegre hotels. **ESPM Magazine. Sao Paulo,** v. 10, p. 39-54, May/June, 2003.

HAYES, B.E. **Measuring Customer Satisfaction.** Rio de Janeiro: Qualitymark, 2001.

HAIR, J.F.Jr.; ANDERSON, R. E.; TATHAN, R. L.; BLACK,W. C. **Multivariate Data Analysis.** 4. ed., Prentice Hall: USA, 1998.

JOHNSON, R.A.; WICHERN, D.W. **Applied multivariate statistical analysis**. 3. ed., New Jersey: Prentice-Hall, 1992.

___. **Applied multivariate statistical analysis**. 4. ed., New Jersey: Prentice-Hall, 1998.

JURAN, J.M. **Juran on quality leadership:** a guide for executives. Sâo Paulo: Pioneira, 1990.

KACHIGAN, K.S. **Multivariate statistical analysis:** a conceptual introduction. New York: Radius Press, 1982.

KAISER, H.F. **The varimax criterion for analytic rotation in factor analysis**. Psychometrica: USA, 1958.

___. The application of electronic computers to factor analysis. **Educational and Psychological Measurement**. v. xx, n. 1, 1960.

KENDALL, M.G. **A course in multivariate analysis.** London: Griffin, 1957.

KOTLER, P. **Marketing Management, analysis, planning, implementation and control.** 4 ed., Sâo Paulo: Atlas, 1994.

___. **Marketing Management.** 4 ed., Sâo Paulo: Atlas, 1996.

LATIF, S.A. Factor Analysis helping to solve a real Marketing Research problem. **Caderno de Pesquisas em Administraçao** *[on line]*. 1994, v. 00, n. 0, [cited 2003-0910]. Available from < http:// www.ead.fea.usp.br/cad-pesq/arquivos/coo-art05>.

LOVELOCK, C.; WRIGHT, L. **Services - Marketing and Management.** Sao Paulo: Saraiva, 2002.

MACEDO, S.G. **Teaching Performance through Student Evaluation:** A methodological proposal to subsidize university management. Florianópolis: UFSC, 2001, 131 p. Thesis (Doctorate in Production Engineering) - Federal University of Santa Catarina, 2001.

MALHOTRA, N. **Marketing Research:** an applied orientation. Porto Alegre: Bookman, 2001.

MARDIA, K.V.; KENT, J.T.; BIBBY, J.M. Multivariate analysis. London: Academic, 1979.

MATTAR, F. N. **Pesquisa de Marketing.** 4 ed., Sao Paulo: Atlas, 1993.

MONTGOMERY, D.C. **Introduction to statistical quality control**. 3. ed., New York: John Wiley & Sons, Inc. 1997.

MORRISON, D.F. **Multivariate Statistical Methods**. 2. ed., New York: McGraw Hill, 1976

PEREIRA, J.C.R. **Anâlise de Dados Qualitativos:** estratégias metodológicas para as ciências da saù, humanas e sociais. 2. ed.; Sao Paulo: USP, 1999.

PLA, L.E. **Multivariate analysis:** principal components method. Venezuela. General Secretariat of the Organization of American States: Washington, D.C., 1986.

SCREMIN, M.A.A. **Method for selecting the number of principal components based on fuzzy logic**. Florianópolis: UFSC, 2003, p. 124 Thesis (Doctorate in Production Engineering) - Federal University of Santa Catarina, 2003.

SOUZA, A. M. **Principal Components:** Application to the reduction of economic variables for the study of time series. Santa Maria: UFSM, 1993, p. 155. Dissertation (Master's Degree in Production Engineering) - Federal University of Santa Maria, 1993.

___. **Monitoring and adjusting feedback in multivariate production processes.**

Florianópolis: UFSC, 2000, 166p. Thesis (Doctorate in Production Engineering) - Federal University of Santa Catarina, 2000.

WERKEMA, M.C.C. **As ferramentas da qualidade no gerenciamento de processos.** Belo Horizonte: Christiano Ottoni Foundation, 1995.

TQM: Everyone wants to improve. Total Quality Management. Sao Paulo: Imam, 1994.

AZEVEDO, J.A.T.; TAKAKI, T. A matemàtica na satisfaçao dos clientes. **Banas Qualidade, Sao**

Paulo, n 132, p. 50-56, May 2003.

ANNEXES

EVALUATING CUSTOMER SATISFACTION AT RBS TV SANTA ROSA

Company Data

1. Name (optional) ____

2. City: ____

3. Type of company ( )

1) Commercial 2) Industrial 3) Service Provider 4) Educational 5)Other

4. Number of employees ( )

5. Company classification

1) Micro 2) Small 3) Medium 4) Large .( )

6. Time in business

1) 1 to 2 years 2) 3 to 4 years 3) 5 to 10 years 4) more than 10 years ( )

7. Public or Private Company

1) Public 2) Private ( )

8. What kind of media does the company use most

1) Radio 2) TV 3) Newspaper 4) More than one media ( )

9. What type of media do you think brings the most return?

1) Radio 2) TV 3) Newspaper ( )

10. How often do you advertise on RBS TV?

1) Monthly 2) Half-yearly 3) Annual 4) No set date ( )

11. You are currently broadcast on RBS-TV

1) Yes 2) No ( )

12. If not, for what reason

13. Annual investment in advertising on RBS-TV ( ) 1) Less of R$ 1.000,00 2) From R$ 1.000,00 to R$ 5.000,00

3) From R$ 5.000,00 to R$ 10.000,00 4) From R$ 10.000,00 to R$ 50.000,00

5) More than R$ 50,000.00 6) Other:____________________

14. It is usually an agent or their company that seeks out RBS-TV's services

1) Agent 2) Company ( )

Please indicate the degree to which you are satisfied or dissatisfied with the following aspects of the service provided to you by RBS TV Santa Rosa. Mark the appropriate item with an X, using the table below:

1. I am Very Dissatisfied with this aspect (MI)
2. I am dissatisfied with this aspect (I)
3. I am neither satisfied nor dissatisfied with this aspect(N)
4. I am satisfied with this aspect (S)
5. I'm very satisfied with this aspect (MS)

Level of Satisfaction

MI I N S MS

1. RBS TV Santa Rosa as a media option.
2. As for the return you get from investing in advertising on RBS TV Santa Rosa.
3. As for alternative ads.
4. The ad proposed by the agency met his needs. ____________________
5. The media hours offered catered for their target audience.

7 . Regarding the agent's service

(advertising sales). ____________________

8 . When you contact the company, you receive a reply to all your requests, complaints and/or suggestions.

9 . When I made an appointment, the agent was available for the meeting at a time that suited me.

___

10 The agent was prompt when I arrived at the meeting. ____________________

11 Punctuality of meeting start times.

12 As for after-sales service. ____________________

13 RBS TV Santa Rosa price list. ____________________

14 Payment terms. ____________________

15 Whether you're happy to continue or return

advertising on RBS TV Santa Rosa.

16 Make a suggestion about the service provided by the Sales Department of RBS TV Santa Rosa.

Printed by Books on Demand GmbH, Norderstedt / Germany